FOSS Science Resources

Plants and Animals

Full Option Science System
Developed at
The Lawrence Hall of Science,
University of California, Berkeley
Published and distributed by
Delta Education,
a member of the School Specialty Family

1487698
978-1-62571-284-4
Printing 5 – 4/2018
Standard Printing, Canton, OH

Table of Contents

What Do Plants Need?

Many people grow **plants**.
They grow plants in pots and window boxes.
They grow plants in gardens and on farms.
Farmers grow plants to sell.
They know the **basic needs** of plants.

Plants need water.
Their **roots** take up water.
Water travels from the roots to the stems,
leaves, and flowers.

Plants need **nutrients**.
Nutrients come from the soil.
Water carries the nutrients through the plants.

Plants need light and **air**.
Their leaves capture light.
Green leaves use **sunlight** and air to make **food**.
Plants need food to grow.

Plants need space.
Roots need space to grow.
Crowded roots can't get enough water and nutrients.
Crowded leaves can't get enough light.

Does this plant have what it needs to grow?

Thinking about What Do Plants Need?

1. What are the basic needs of plants?

2. What part of a plant takes up water?

3. Where do nutrients come from?

4. What part of a plant captures light?

The Story of Wheat

People use plants in many ways.
They make clothing and houses from plants.
They also use plants for food.

Wheat is an important food plant.
People use wheat to make **flour**.
Flour is used in cooking and baking.

How does wheat grow?
First, farmers **sow** wheat seeds
in big fields.

Soon, the seeds sprout.
The wheat looks like grass.
The plants grow bigger and bigger.
Each wheat plant grows new seeds.
The seeds are called **grain**.

The wheat plants dry and turn golden.
It's time to **harvest** the wheat grain.
The farmer drives a machine called a **combine** over the field.
The combine cuts the wheat plant.
The combine collects only the grain.

Farmers store the grain in large silos.
Later, it will go to a mill.

But farmers keep some of the grain.
Why do you think they keep some of it?

The grain at the mill is ground into flour.
The flour flows into sacks.
Bakeries and grocery stores buy the flour.

Cooks and bakers use flour.
They mix it with water and other things.
The mixture might be baked in an oven.
It might be cooked on a stove.
When it's done, there is always something good to eat!

Bread, pasta, and tortillas are made from flour.
Can you think of other things made from flour?

Thinking about The Story of Wheat

1. What part of the wheat plant is grain?

2. How does grain become flour?

3. Why do farmers save some of the grain?

4. What foods are made from flour?

Variation

Variation means difference.
Look at the flowers on this page.
They are not all the same.
Do other plants and **animals**
have variation?

People are different.
Some people are short.
Some have brown eyes.
Some have black hair.
Some have freckles.
Everyone is different when you look closely enough.

These black bears have the same shape and size.
But like other animals, they have variation too.
What kind of variation do you see?

Trout have color variation, too.
Some of them are silver.
Others are brightly colored.
Some have a lot of spots.
Others have only a few spots.
Trout have both color and pattern variation.

These shells are all from the same kind of scallop.
What kinds of variation can you see?

Twelve students brought pets to school
for show and tell.
The pets are all the same kind of animal.
It is easy to see that the dogs are different sizes.
What other variations can you observe?

Here is a garden with marigolds.
Do you see any variation?

Thinking about Variation

1. Name five variations you can observe about people.

2. Tell about variations in trout.

3. Think of another animal, and tell about its variations.

What Do Animals Need?

Animals live in different **habitats**.
Some animals live in water.
Others live on land.
Some animals live on other animals.
But all animals have the same basic needs.

Animals need food.
They eat plants and animals.
Animals that eat plants are **herbivores**.
Animals that eat other animals are **carnivores**.

Animals need water.
Most land animals drink fresh water.
Some animals get water only from their food.
Land animals need air.
Air contains **oxygen** that animals need to live.

Animals need **shelter**.
Shelter protects animals from weather
and other animals.
Burrows and **nests** are safe places for
young animals.

A **terrarium** is a small habitat made in a container. Worms, snails, insects, and plants can live together in a habitat.

Animals are **living** things.
Plants are living things, too.
All living things have basic needs.
What basic needs does a terrarium provide
for plants and animals?

Thinking about What Do Animals Need?

1. What are the basic needs of animals?

2. Why do animals need shelter?

3. Look at the terrarium on page 32. How is it like the one that you made in class?

4. Why do animals need plants?

Plants and Animals around the World

Plants and animals live in many different habitats.
The ocean provides many places for plants and animals to live.
What do you see in this saltwater habitat?

Many plants live in the **rain forest**.
Some rain forest plants are very tall.
They grow tall to get the light they need.
Short rain forest plants have big leaves.
Big leaves are needed to collect enough
light in a shady rain forest.

Frogs and sloths live in rain forests.
This green frog is hard to see on the green leaves.
It catches insects with its long, sticky tongue.
Water rolls right off the frog's smooth skin.
Frogs lay eggs on wet leaves.

Sloths have strong legs and claws.
They move slowly in the tops of trees.
They eat leaves from the trees.
The baby sloth holds onto its mother.
The mother sloth and her baby are safe
high in the trees.

Plants grow on the cold **tundra**.

Summer days are long.

Tundra plants grow flowers and make seeds in summer.

Winter days are short.

Tundra plants stop growing in winter.

Caribou and lemmings live on the tundra.
Caribou eat the short tundra plants
during summer.
They drink from the many rivers.
The caribou's thick fur helps it stay warm.
Before winter, caribou travel to a warmer place.

Lemmings stay on the tundra all year.
In summer, they eat and store seeds.
They make nests with dry grass.
Their nests and thick fur keep them warm
in winter.
Fleas on the lemmings stay warm, too.

Plants grow in hot, dry, windy **deserts**.
Desert plants get lots of light.
But they get very little water.

Cactus plants have long roots.
The roots spread in the desert soil.
When it rains, the roots take up water.
Cactus plants store water in their thick stems.

Lizards and elf owls live in the desert.
Lizards **thrive** in the hot Sun.
They eat insects.
Lizards get water from their food.
Lizards have big feet with sharp claws.
They can run quickly on sand.
They climb on rocks to escape from
other animals.

Elf owls do not thrive in the hot Sun.
This elf owl uses a cactus for shelter.
Elf owls hunt insects in the cool nights.
Big eyes help owls see in the dark.

Lots of plants grow in **forests**.
Summers are hot.
Winters are cold.
Many forest trees and bushes
lose their leaves in fall.

It rains a lot in summer.
It snows in winter.
There is always water for plants and animals in the forest.

Blue jays and chipmunks live in the forest.
Blue jays can fly all over the forest.
They fly up into trees to find insects.
They fly down to streams for water.
Trees provide shelter for blue jays.

Chipmunks live in the forest all year.
In fall, they gather nuts.
They store the nuts in their burrows.
Chipmunks stay in their burrows during winter.
Their warm fur and stored nuts help them **survive**.

Grasslands have lots of grass.
But they often don't have trees.
Summers are warm and sunny.
Winters are cold.
Rain and snow provide water
for plants and animals.

In fall, the grass dies and turns golden.
Sometimes, wildfires burn the dry grassland.
Fire kills young trees and bushes in the grassland.
Now there is more space for new plants.
Plants like grasses thrive after a fire.

Prairie dogs and hawks live in grasslands.
Prairie dogs live in tunnels under the ground.
They use their strong, sharp claws to dig tunnels.
They come out to eat grass seeds and stems.
They are always watching for danger.

Hawks soar over the grassland on broad wings.
They catch small animals with their strong, sharp talons.
Hawks thrive when they catch and eat prairie dogs.
Hawks are **predators**.

Water lilies live in ponds.
Their roots grow in the mud at the bottom of
a pond.
In spring, water lilies grow big leaves called pads.
The pads rest on the surface and collect sunlight.
If a pond freezes, the leaves die.
Dead leaves feed the animals living under the ice.

Frogs and perch live in ponds.
Frogs can live in water.
They can also live out of water.
Sometimes they sit on lily pads.
They are waiting for insects to eat.
Frogs use their long, sticky tongue to catch food.
If surprised, frogs leap into the water.
They can swim quickly with their strong legs.

Perch can find shelter under lily pads.
Their colors and patterns make them hard to see.
Perch wait in the shadows for insects to eat.
If surprised, perch swim to safety.
They move quickly using a broad tail and fins.

Thinking about Plants and Animals around the World

Plants and animals live in many different habitats.
Habitats include the ocean, rain forest, tundra, desert, forest, grassland, and pond.
How do animals get food and shelter in their habitats?

Learning from Nature

Many animals have **structures** that help them swim in water.
Ducks have structures that help them swim.
What structures can you see that help ducks move in water?

Frogs have structures that help them swim. What structures can you see that help frogs move in water?

Ducks and frogs have big webbed feet.
Webbed feet help them push through the water.
What can people learn about swimming from frogs and ducks?

An engineer designed some big webbed feet for people.
The big webbed feet are called flippers.
People wearing flippers can move better in water.

Have you ever seen a squirrel climb a tree?
Squirrels are good climbers.
Bears are good climbers, too.
How can they climb like that?

What structures do you see that help squirrels and bears climb?
Squirrels and bears have strong sharp claws.
What can people learn about climbing trees from squirrels and bears?

Sometimes people need to climb smooth wooden poles.
People don't have strong sharp claws for climbing.
An engineer designed a tool for climbing poles.
The engineer made a pair of boots with big sharp climbing spurs.

Each spur on the boot is like a single strong claw.
A worker wearing these boots can climb to the top of a pole.
A tree trimmer wearing these boots can climb a tree.

Seals live in cold ocean water.
Seals are mammals like people and must stay warm.
How are seals able to live in such cold water?
Seals have a layer of fat, called blubber,
under their skin.
The blubber is **insulation** and acts like a blanket.
The blubber keeps the heat from leaving
the seal's body.

This is what the air spaces in the fabric look like.

Sometimes people need to work in cold water.
What can people learn from seals about surviving in cold water?

Engineers developed a rubber fabric called neoprene.
Neoprene is filled with lots of tiny air bubbles.
The fabric acts like a layer of blubber and is good insulation.

Neoprene can be glued into tight-fitting suits called wetsuits.
Divers wearing a layer of neoprene insulation can swim and work in very cold water.

Some land animals can live in very cold weather.
Animals like arctic foxes and polar bears have thick fur coats.
People looked closely at the fur of arctic foxes.
They saw a layer of very fine fluffy fur.

Snowy owls can also live in cold places.
People studied the feathers of snowy owls.
They saw a layer of very fine fluffy feathers.

What did people learn about staying warm from arctic animals?
A layer of fluffy fur or feathers is good insulation.
Cold air can't easily get through a fine fluffy layer.
The fine fluff keeps warm air near the animal's body.

Engineers used this information.
They developed a fluffy material
made from fine polyester threads.
The fluffy polyester is good insulation.
People use polyester fluff to stuff coats.
It keeps people warm on the coldest days.

Animals and Their Young

Animals live and grow in many different habitats.
They get food, water, and shelter from their habitats.
They might find shelter in a nest or hole.
Animals can move around to get food, water,
and shelter.

Many young animals need care.
Baby monkeys and hummingbirds
cannot feed or clean themselves.
Parents feed and clean their young.

This mother lion is cleaning her cub.
A young fox is being groomed, too.

Some **offspring** get their first food from their parents.
This young penguin is getting some seafood.
Can you see the camel getting milk from its mother?

As the young get older, some parents teach their offspring how to get food. Mother grizzly bears teach their cubs to catch fish.

The cub might have to try many times before catching a fish.

Young ospreys need a safe place to grow.
Ospreys build large, strong nests in high places.
The eggs and baby birds are safe from predators.

Some spiders shelter their eggs in a strong silk case.
What other shelters for young do you see?

Some animals hide their babies to keep them safe.
A hole in a tree is a good hiding place for baby woodpeckers.
Some birds hide their nests in dense bushes.

Some parents carry their offspring to keep them safe.
What animals do you see carrying their babies?
If there is danger, the parent can move quickly to a safer place.

Most young animals need to stay warm.
Some mothers keep their babies close.
Heat from the mother's body keeps the babies warm.
How do other animals keep their young warm?

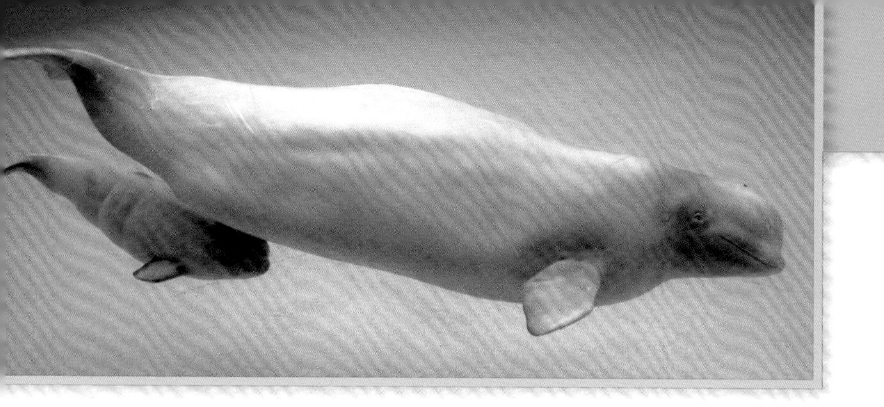

Young animals learn to walk, climb, and swim like their parents.
Moving quickly can help young animals survive.
Parents help their young stay away from danger.

Parents care for their offspring in many ways.
They make a safe shelter.
They teach their young to feed, climb, and swim.
Some offspring stay close to their parents.
How are these parents caring for their young?

Thinking about Animals and Their Young

1. How do parents care for their offspring?

2. How do parents keep their young warm?

3. Tell how parents provide food for their offspring.

4. What do young animals do if there is danger?

Glossary

air a mixture of gases that and animals and plants need o live **(6)**

animal a living thing that is not a plant **(19)**

basic need something that plants and animals need to urvive, such as air, water, ood, space, light, and shelter. **3)**

carnivore an animal that ats only other animals for ood **(28)**

combine a machine that cuts wheat and separates the grain **13)**

desert a dry place with little ain **(41)**

flour a fine powder made rom grinding wheat seeds **11)**

food what plants and animals need to survive **(6)**

forest a place with many trees and other plants **(45)**

grain a hard seed that grows on a wheat plant **(12)**

grassland a place with a lot of grass and often no trees **(49)**

habitat the place or natural area where plants and animals live **(27)**

harvest to gather a crop, such as wheat **(13)**

herbivore an animal that eats only plants for food **(28)**

insulation a layer of material that prevents the movement of heat **(65)**

leaf a structure on a plant that is usually green and makes food from sunlight **(4)**

living alive. All living things have basic needs and produce offspring. **(32)**

nest a safe place where animals live and raise their young **(30)**

nutrient what living things need to grow and stay healthy **(5)**

offspring a new plant or animal produced by a parent **(74)**

oxygen a gas in air and water that plants and animals need to live **(29)**

plant a living thing that has roots, stems, and leaves. Plants make their own food. **(3)**

predator an animal that hunts and catches other animals for food **(52)**

rain forest a warm, wet place with many trees and other plants **(35)**

root a part of a plant that grows in soil **(4)**

shelter a safe place where animals live. A shelter protec an animal from weather or other animals. **(30)**

sow to plant a seed **(11)**

structure any identifiable part of a plant or animal **(57**

sunlight something plants need to make food **(6)**

survive to stay alive **(48)**

terrarium a small containe with soil where plants and animals can live **(31)**

thrive to grow fast and stay healthy **(43)**

tundra a place in the arctic high on mountains **(38)**

variation difference **(19)**

wheat a type of grass that makes seeds that can be ground into flour **(11)**